SharePoint 2013 Installation

Step by Step

Alan Richards

DEDICATION

This first book in my technical writing career must be dedicated to my beautiful wife. Without her ongoing and steadfast support I would not be where I am today. It's a cliché I know, but she is my rock and inspiration in everything I choose to do.

Contents

Introduction

With the release of SharePoint 2013 comes a whole new raft of technologies to understand and implement.

This book will go through the setting up of SharePoint 2013 itself from creating the setup accounts to the actual install of the software. We won't be covering the installation of SQL Server in this book, we assumes you have a functioning SQL Server ready for the installation of SharePoint 2013, for all examples an instance of SQL 2012 server is used.

The book will also cover the setup and implementation of the new Microsoft Workflow Manager and Office Web Apps which now run as separate servers and are accessed by SharePoint 2013.

This book uses a very simple infrastructure model which is ideal for demonstrating the techniques for installing SharePoint 2013 and its associated servers, but this model should not be used in a production environment.

Infrastructure

For this book we are going to use a simple SharePoint 2013 infrastructure model which as mentioned should not be used in a production environment.

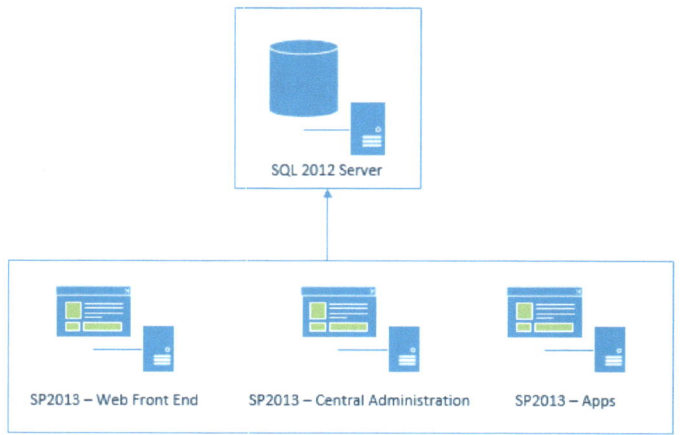

This topology consists of the following roles

SQL 2012 Server

Database server for all SharePoint 2013 databases

SP2013 - Central Administration

SharePoint 2013 server that will host the central administration service

SP2013 - Web Front End

SharePoint 2013 server that will host the web service for all users

SP2013 – Apps

A standalone server that will not have SharePoint 2013 installed but will be used to host the Workflow Manager Server and Office Web Apps Server

Security Accounts

As with SharePoint 2010 a policy of least privilege should be applied to the installation of SharePoint 2013, Microsoft recommends certain accounts be created for use with SharePoint 2013. A TechNet article covering the creation of these account can be found here http://technet.microsoft.com/en-us/library/cc678863.aspx

A simplified version of the article can be found below.

Setup user administrator account

This account is used to set up each server in your farm by running the SharePoint Configuration Wizard, the initial Farm Creation Wizard, and Windows PowerShell. It should be the account you use to logon to your SharePoint 2013 servers when you need to carry out any configuration.

The setup user administrator account should be a member of the **Domain Users** account only in active directory but should be a member of the **Local Administrators** group on each SharePoint 2013 server

This account will obviously require access to the SQL 2012 server so a login will need to be created in the SQL 2012 console, it should be created with the **security admin** and **database creator** roles

Additional permissions on the SharePoint 2013 databases and on the SharePoint 2013 servers will be assigned automatically during the configuration of SharePoint 2013

SharePoint farm service account

The server farm account, which is also referred to as the database access account, is used as the application pool identity for Central Administration and as the process account for the SharePoint Foundation 2013 Timer service.

The server farm account should be a member of the **Domain Users** account only in active directory but should not be added to any local groups on the SharePoint 2013 servers

Additional permissions on the SharePoint 2013 databases and on the SharePoint 2013 servers will be assigned automatically during the configuration of SharePoint 2013

Application pool account

The application pool account is used for application pool identity and should be a member of the **Domain Users** group only in active directory, it should not be added to any local groups on the SharePoint 2013 servers.

Additional permissions on the SharePoint 2013 databases and on the SharePoint 2013 servers will be assigned automatically during the configuration of SharePoint 2013

Active Directory Accounts

Following the guidelines that Microsoft publish the following accounts are setup in Active Directory

SP13-Setup – Domain User and Local Administrator on SharePoint 2013 servers

SP13-FarmAd – Domain User

SP13-AppPool – Domain User

Installation Of SharePoint 2013

The three servers that have been prepared for the installation all have Windows Server 2012 installed, updated and connected to the local domain.

The installation for all the SharePoint 2013 servers follows the same process as below but the **Central Administration** server needs to be installed first before any **Web Front End**

Logon to the server as a domain administrator and add the **SP13-Setup** user to the local administrators group

Log off and logon again as the **SP13-Setup** user

In earlier versions of SharePoint 2013 there was a little bug in setup when running on Windows Server 2012 in that it would fail every time while trying to configure the web server application role when installing the pre-requisites

The workaround for this was to install the web server role manually and also install the .NET Framework 3.5 manually also, I have left the instructions for this process in this version for reference and in case you hit the problem at some point in the future

Web Server Application Role Installation

Load up **Server Manager**

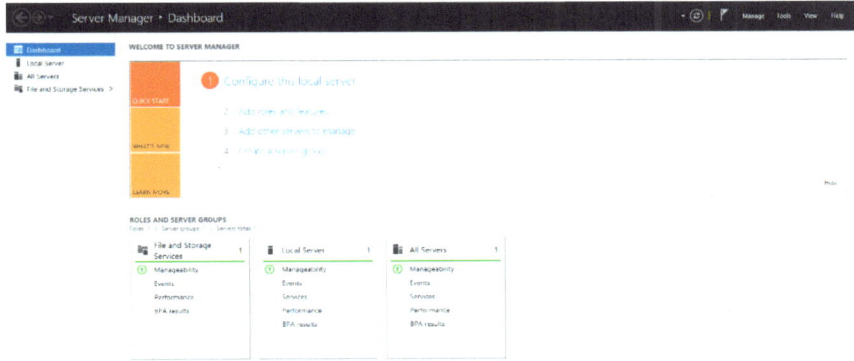

Click on **Manage** and select **Add Roles & Features**

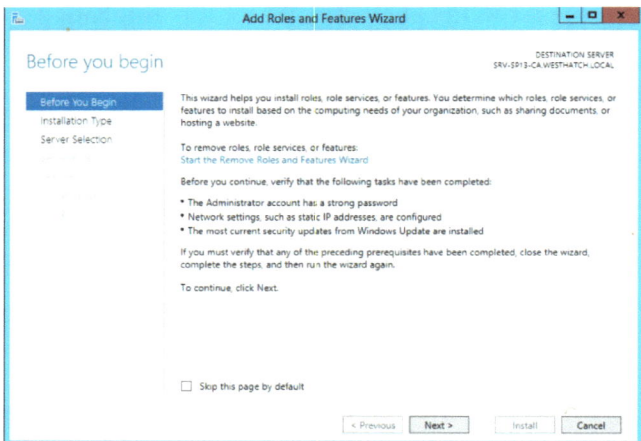

Click **Next** 3 times and select **Web Server (IIS)** from the roles screen

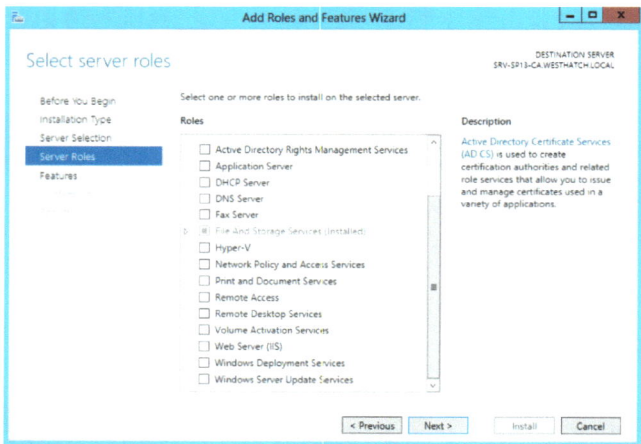

At this point you will be prompted to add features needed by the Web Server role, click **Add Features** to continue

Click **Next** 4 times and then click **Install** on the final screen

Once installation has completed click on **Close**

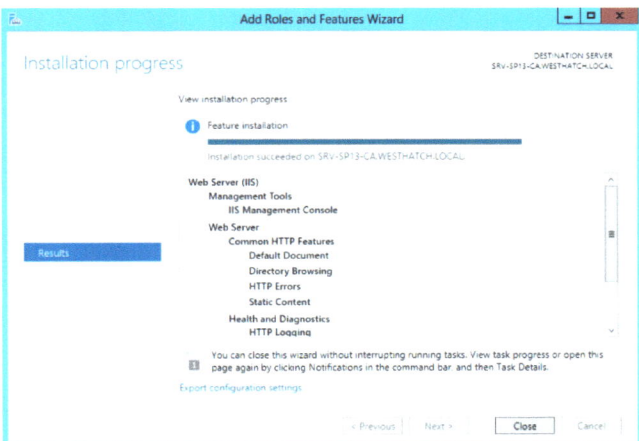

.NET Framework 3.5 Features Installation

Click on **Manage** and select **Add Roles & Features**

Click Next 4 times and select **.NET Framework 3.5 features**

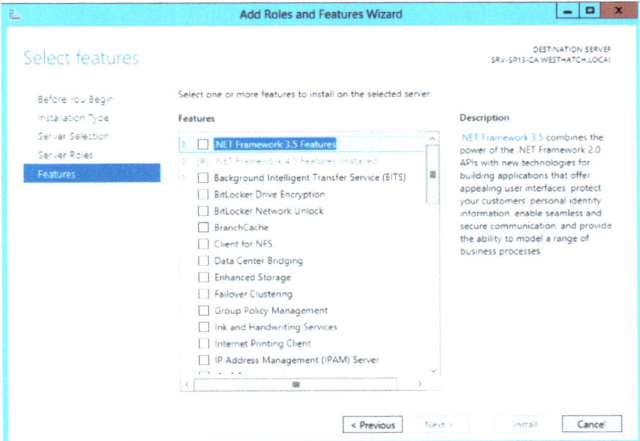

7

Click **Next**

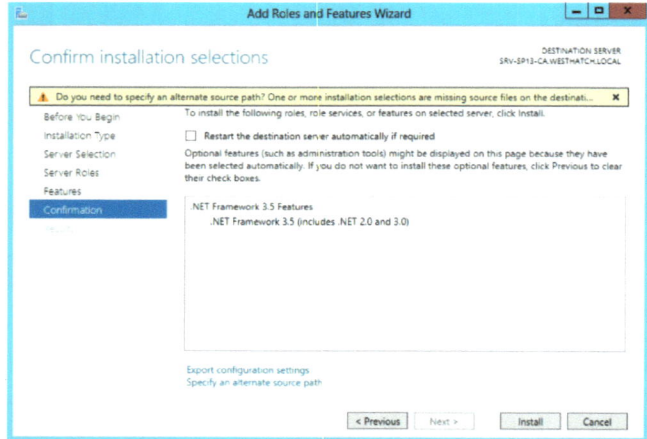

At this screen click on **Specify and alternative source path** (ensure you have the Windows Server 2012 DVD in the DVD drive)

Enter the following path **D:\Sources\SxS** (where D: is the DVD drive on the SharePoint server) and click **OK**

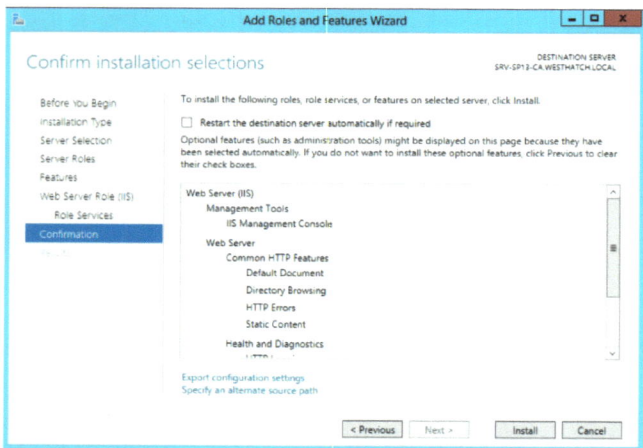

Click **Install**

Once installation has completed click on **Close**

SharePoint 2013 Prerequisites

Now that we have installed the above roles and features we can start with the installation of the prerequisites for SharePoint 2013

Load the SharePoint 2013 DVD and select **Install software prerequisites**

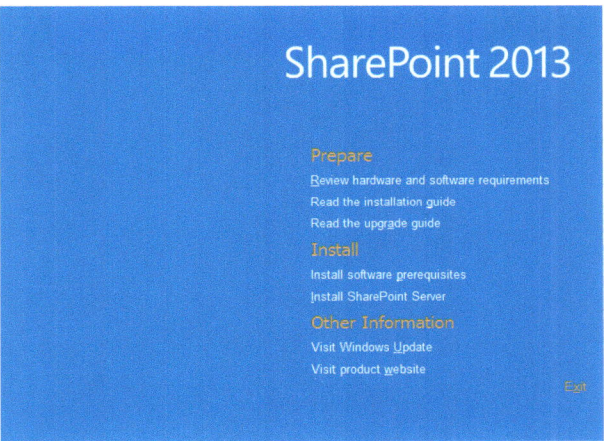

Click **Next** at the welcome screen

Accept the **License Agreement** and click **Next**

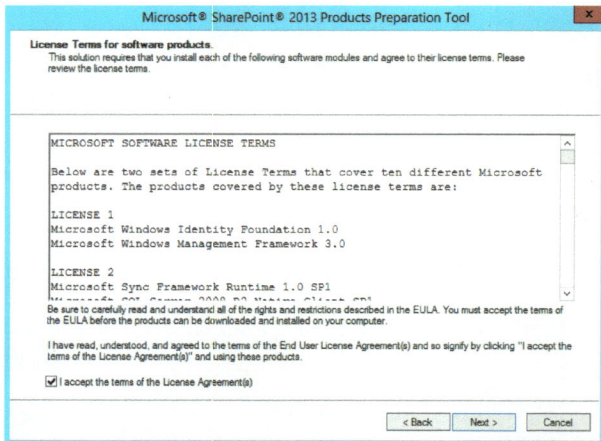

The prerequisites will start to be installed

The server will require a number of reboots before all of the prerequisites are installed and you see this screen

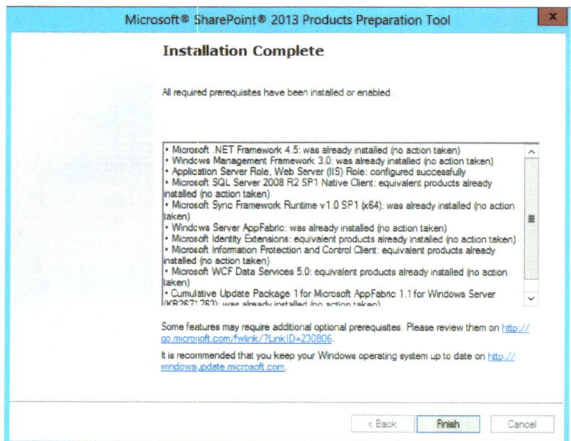

Install Core Software

With the prerequisites installed we can now install the core files for SharePoint 2013. With the SharePoint 2013 Setup User logged in, load the SharePoint 2013 DVD and select **Install SharePoint Server**

Enter the product key and click **Continue**

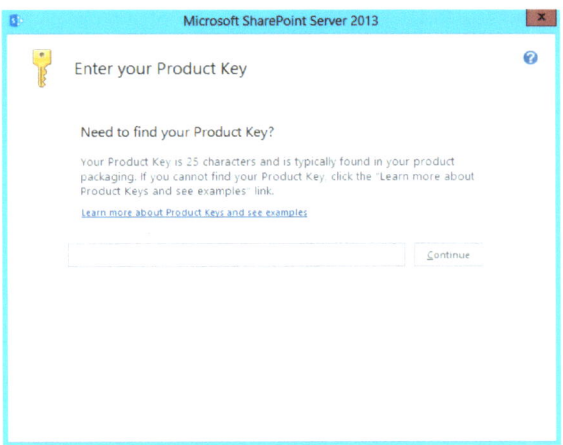

Accept the license terms and click **Continue**

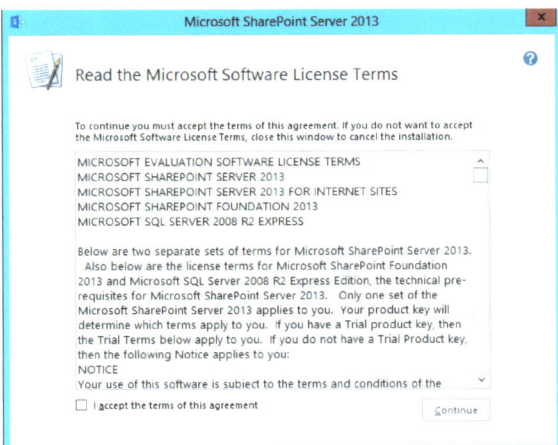

As this is a production environment and we have access to SQL 2012 Server, select **Complete** and also you may want to alter the file locations if necessary for your environment

Once all your selections are made click on **Install Now**

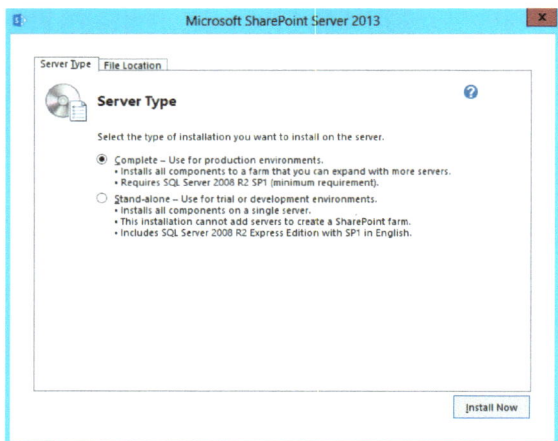

SharePoint Products Configuration Wizard

Once install is complete run the configuration wizard and at the welcome screen click on **Next**

At the prompt click **Yes** to allow services to be restarted

Click on **Create a new server farm** and click **Next**

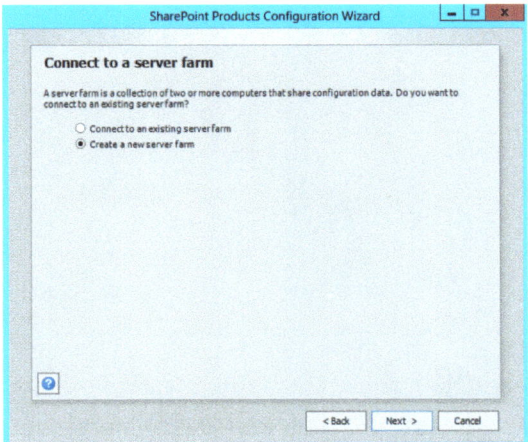

We now need to enter the name of the database server, database name and the user details for the database access account (this should be the **SharePoint Farm Service Account**) and click **Next**

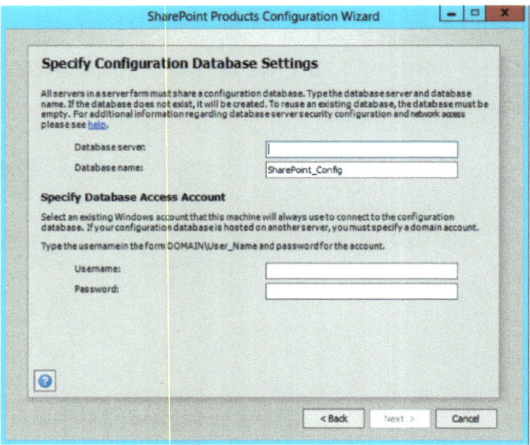

We now need to enter a passphrase, this will be used as a level of security when adding servers to the SharePoint 2013 farm

Enter a complex passphrase and click **Next**

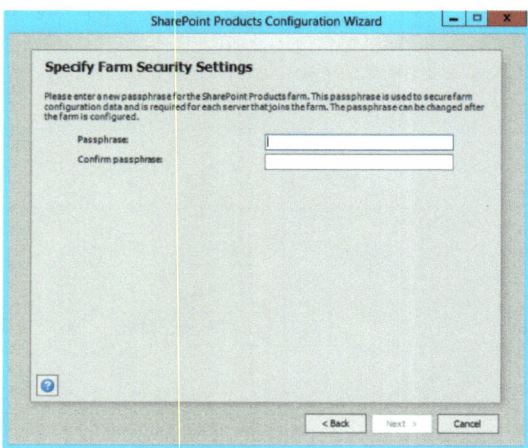

At the next screen select the authentication method and alter the **Central Administration Web Application** port if necessary and click **Next**

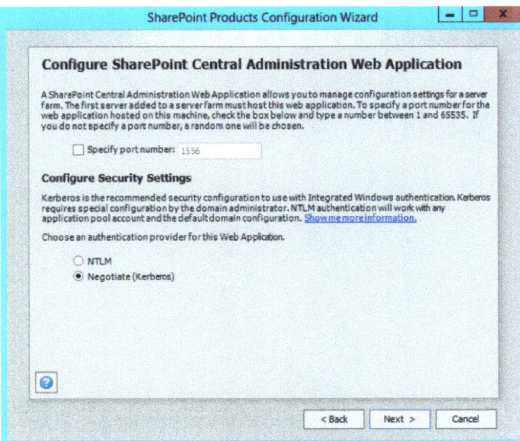

At the summary screen check all the details and click **Next**

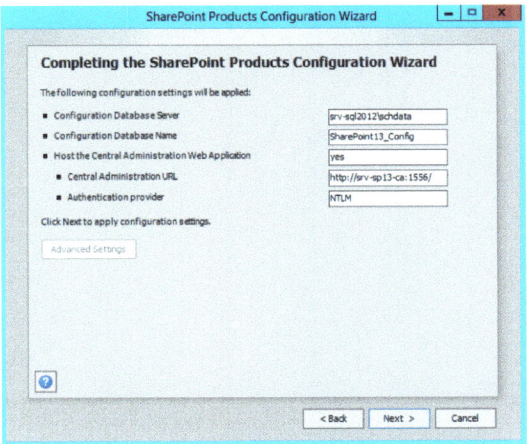

The wizard will now create the configuration database and install all the files needed by SharePoint 2013 and when its finishes you should see this screen

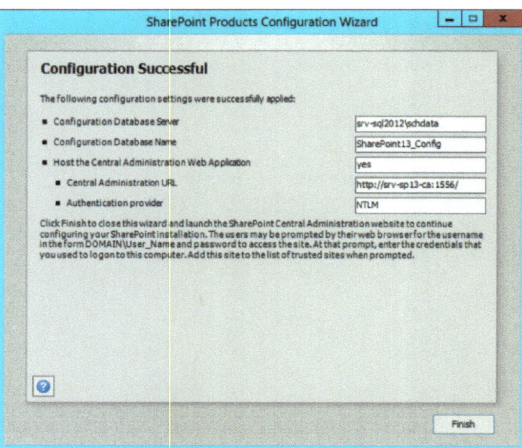

Clicking **Finish** will close this screen and launch the **Central Administration Web Application** in your browser

Adding Servers To A SharePoint 2013 Farm

To add additional servers to the SharePoint 2013 farm we need to follow the same installation procedure as the first server in the farm until we get to the section in the configuration wizard concerned with creating or connecting to a SharePoint 2013 farm

At this screen select **Connect to an existing server farm**

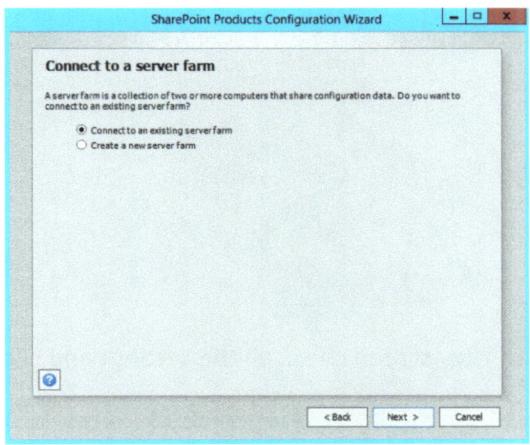

Enter the name of the database server and click on **Retrieve Database Names**

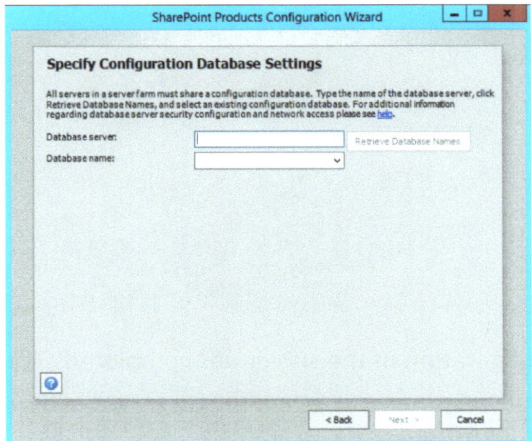

From the drop down list of databases select your SharePoint 2013

configuration database and click **Next**

We now need to enter the passphrase we created earlier to connect this new server to the SharePoint 2013 farm and click **Next**

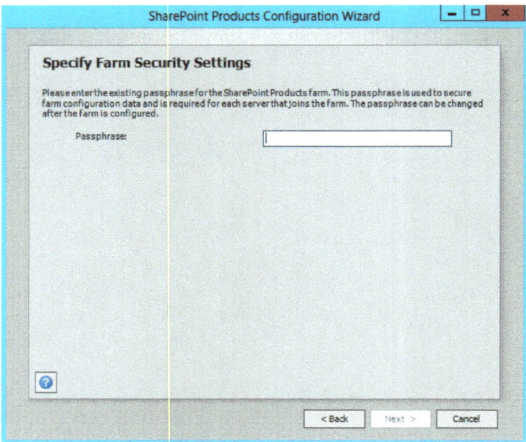

At the summary screen check all the settings and click **Next** to proceed

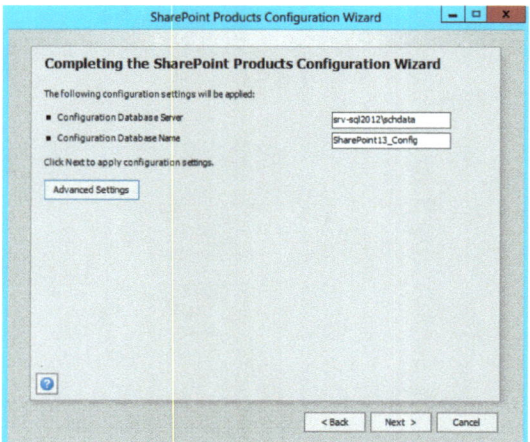

Once configuration of the server has completed click **Finish** to close the screen, we now have an additional server added to the SharePoint 2013 farm

Creating A Web Application

Now that our SharePoint 2013 server farm is complete we are going to add a web application, but the first task will be to create a managed account that will be used for all the application pools

Create New Managed Account

A managed account allows SharePoint 2013 to control aspects of the account and synchronise password changes with Active Directory if necessary

To create a new managed account from the new web application screen select the link in the Application Pool section

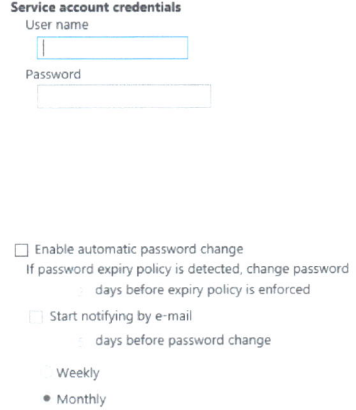

Enter the details of the account with the username in the following format **Domain\Username** and click **OK**

You can now use this account to run your application pools

Create Web Application

Logon to the SharePoint 2013 Server as the SharePoint 2013 Setup Account and launch **Central Administration**

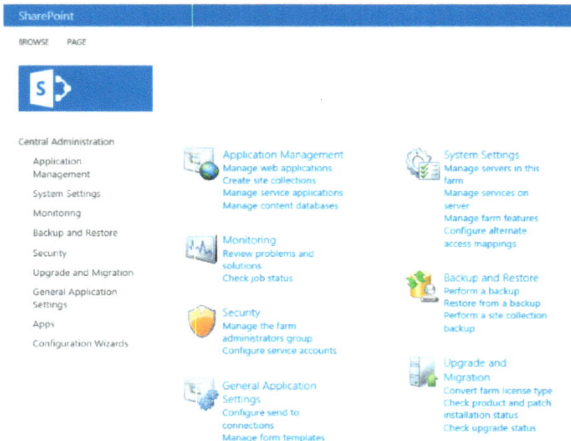

Select **Manage web applications** and select the **New** icon from the ribbon

The next screen contains all the options for the new web application

The key options here are:

Name for the new website – Make it something recognisable

Port for the website – The default is 80, you can use something different if you need to

Do you require anonymous access – Is it going to be a public site

Secure Sockets Layer – If you require the web application to use SSL then all servers will require valid certificates

Authentication method – NTLM or Kerberos

URL of the website – The URL that users will use to access the site (FQDN if using SSL)

Application pool name – Make it something recognisable

Security account – Use the **Application Pool Account** (you should have created a managed account using the instructions above)

Database server – The name of the database server storing the SharePoint 2013 databases

Database name – The name given to the database for this web application

Authentication – The default is Windows and this is the recommended way of accessing the database server

Now we are happy with all of our settings select **OK**

Once the new web application is created this screen will appear, select **OK** to close it

Application Created ✕

The Microsoft SharePoint Foundation Web application has been created.

If this is the first time that you have used this application pool with a SharePoint Web application, you must wait until the Internet Information Services (IIS) Web site has been created on all servers. By default, no new SharePoint site collections are created with the Web application. If you have just created a Forms Based Authentication (FBA) Web application, then before creating a new site collection, you will need to perform some additional configuration steps.

Learn about how to configure a Web application for FBA.

Once you are finished, to create a new site collection, go to the Create Site Collection page.

OK

Create Site Collection

We now have a web application created and need to create a site collection that users can access

From the main central administration page select **Create site collections**

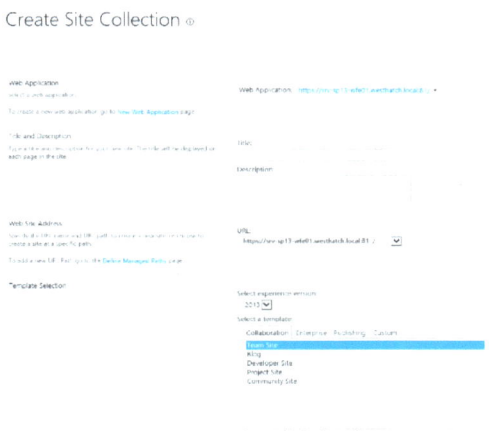

We only have one web application so this should be selected by default at the top of the screen

Enter a title and description for the site collection

We also need to decide if the site collection will sit at the root of the web application or if it will sit under the sites sub domain, as this is the main site collection for this web application it will sit at the root of the web application

You can also select the experience version (2010 or 2013) and the type of site, in this case we have opted for the 2013 experience and a team site

We also need to specify a primary & secondary site collection administrator, in this case we will use the SharePoint 2013 Setup as the primary and Farm Service Account as the secondary

Once all the selections are made select **OK** to create the site collection

Installing Office Web Apps

New for SharePoint 2013 is the introduction of Office Web Apps as a separate server system that can't coexist with a SharePoint 2013 installation

In this books scenario we have a separate server that we will use to install Office Web Apps to, this installation of Office Web Apps is then connected to SharePoint so that it can be used for opening and editing documents and also used by SharePoint 2013 to show previews of documents to users from within document libraries and also in search results

The Office Web Apps installation file can be downloaded from the Microsoft download centre (http://www.microsoft.com/en-us/download/details.aspx?id=35489)

Prerequisites

Office Web Apps server requires a number of prerequisites to be installed prior to running the installation file and the simplest way to install them is to run a PowerShell script. As this example is using Windows Server 2012 as its operating system the PowerShell script is shown below.

Add-WindowsFeature Web-Server,Web-Mgmt-Tools,Web-Mgmt-Console,Web-WebServer,Web-Common-Http,Web-Default-Doc,Web-Static-Content,Web-Performance,Web-Stat-Compression,Web-Dyn-Compression,Web-Security,Web-Filtering,Web-Windows-Auth,Web-App-Dev,Web-Net-Ext45,Web-Asp-Net45,Web-ISAPI-Ext,Web-ISAPI-Filter,Web-Includes,InkandHandwritingServices

---

You may need to restart after the installation of the prerequisites

Installation

As we are using Windows Server 2012 as the operating system the Office Web Apps disc image file can be opened directly from within Windows explorer, if you are using another Windows operating system then you will need to write the image file to a DVD

Run the Setup.exe file and select a location for the Office Web Apps server files and select **Install Now**

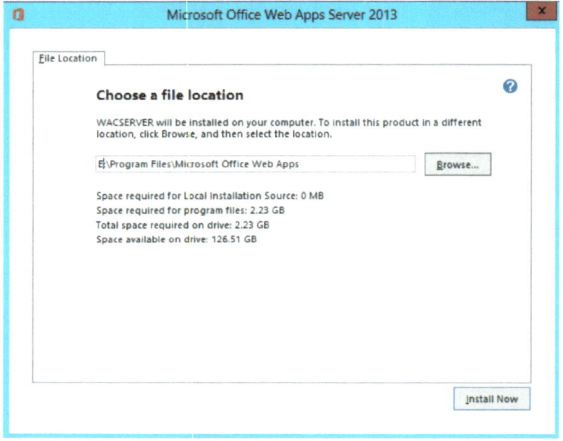

Certificates

Once installation is complete we need to create certificates that will be used by Office Web Apps Server to enable secure communication

The use of certificates is recommended for production environments and a requirement if you want to use Office Web Apps as an external resource

The certificates will need to be from a trusted source, which could be your internal certificate authority if you plan on using them internally only or an external certificate authority for external use

The certificates need to be installed to your IIS installation on the Office Web Apps Server

Configuration

Once the certificates are in place a simple PowerShell command is used to create the Office Web Apps server farm

New-OfficeWebAppsFarm -InternalUrl https://yourserver.local - ExternalUrl https://webapp.yourdomain.com –CertificateName "OfficeWebApps Certificate" –EditingEnabled

The certificate name is the friendly name of the certificate applied to the Office Web Apps server and the –EditingEnabled command allows the editing of documents by Office Web Apps

You should only enable editing if you are licensed to do so

To verify that the Office Web Apps server is functioning navigate to https://yourserver.local/hosting/discovery where you should see a screen of xml if all is working

Claims Based Authentication

For SharePoint 2013 to access Office Web Apps the web application must use claims based authentication. If the web application uses classic mode authentication then running a PowerShell script will convert it to claims based authentication

---

Convert-SPWebApplication -Identity
"http://yourwebapplication:port" -To Claims –
RetainPermissions

Licensing

To enable users to edit documents using Office Web Apps they need to be assigned licenses to edit.

This is achieved by running a series of PowerShell commands on the SharePoint 2013 Central Administration server

---

Get-SPUserLicense

$x = New-SPUserLicenseMapping -SecurityGroup <ADsecuritygroup> –License OfficeWebAppsEdit

$x | Add-SPUserLicenseMapping

Enable-SPUserLicensing

$x is a variable that holds the mapping object to input at the **Add** command

<ADsecuritygroup> is the AD group you want to assign the license to

Bind Office Web Apps

We now need to bind SharePoint 2013 to the Office Web Apps server, again this is achieved by running a PowerShell command on the SharePoint 2013 Central Administration server

---

New-SPWOPIBinding -ServerName <WebAppServerName>

This will by default use HTTPS so the FQDN of the Office Web Apps server needs to be used

Office Web Apps In Use

To test the functionality of Office Web Apps upload a file to a document library and click on the file from the SharePoint 2013 site, this should load the file in the corresponding web app

Office Web Apps will also be used by the preview service in document libraries and search once a full search crawl has taken place, however I will not be covering the setting up of search with SharePoint 2013

Workflow Manager

With SharePoint 2013 comes a new way of doing workflows

Installing SharePoint Server 2013 gives you the same functionality as a SharePoint 2010 workflow out of the box, however to make use of the much greater feature set that is workflows in SharePoint 2013 you need to install the new Workflow Manager

Co-located or Standalone

Before installing you need to decide if you are going to run the Workflow Manager co-located with a running SharePoint 2013 instance or if you are going to install it as a standalone server. Installing it as a standalone server has obvious advantages in that it removes the load of running workflows from your SharePoint farm to a server that you can tailor to your specific needs

Installing Workflow Manager

The installation of Workflow Manager uses the Web Platform Installer from Microsoft and can be downloaded from here http://go.microsoft.com/fwlink/?LinkID=252092

Once the installer runs then it will load up the Web Platform Installer

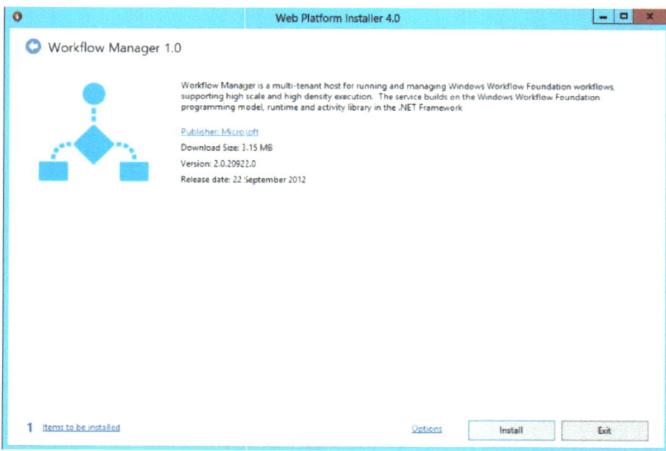

Clicking **Install** will pop up a window showing you the pre-requisites

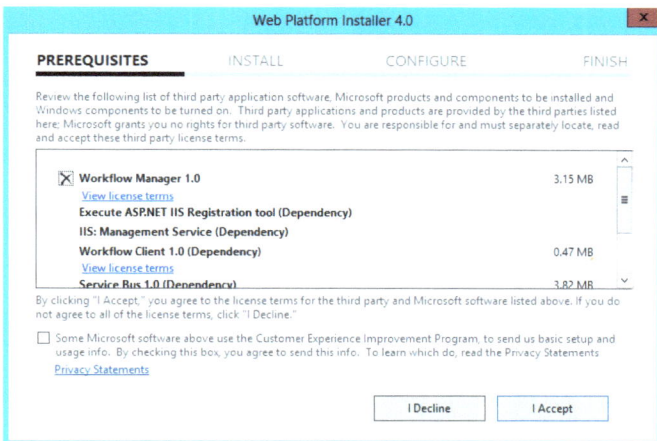

Clicking **I Accept** will start the install of the pre-requisites

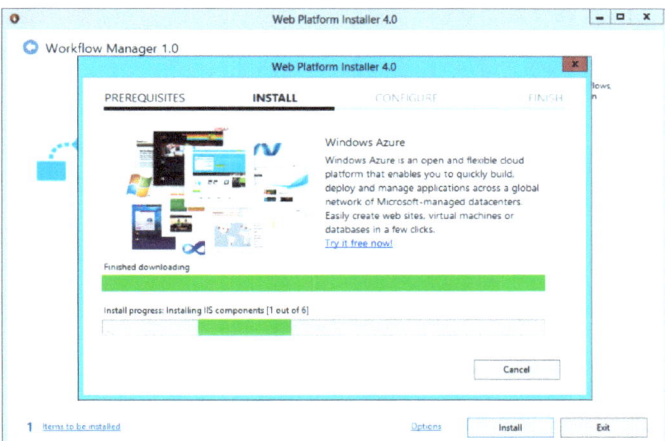

Configuring Workflow Manager

Once installation has completed you will see the configuration window

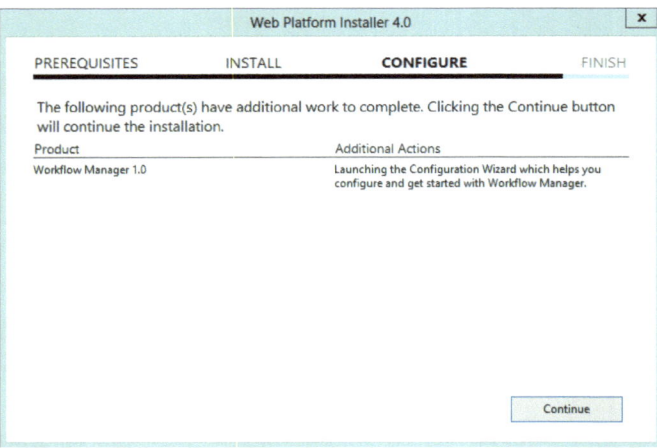

Clicking **Continue** will open the configuration screen

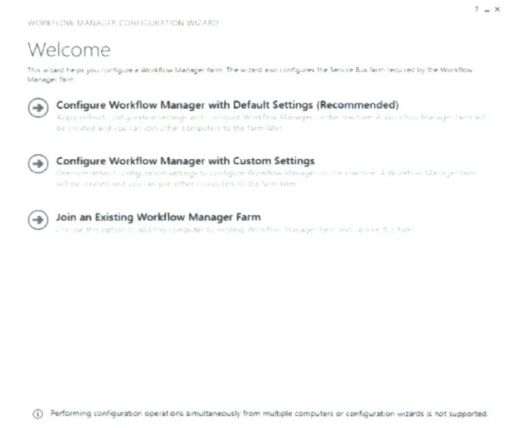

From this screen you can choose to configure with the default settings, this will install a standard Workflow Manager Farm which you can join servers to in the future or you can choose to select the option to use custom settings

For this example I am going to go through the custom settings option so you can see what types of settings you can choose

Connecting To SQL Server

The first screen is where you set the database connection and database names. You also use this screen for authentication and certificate settings

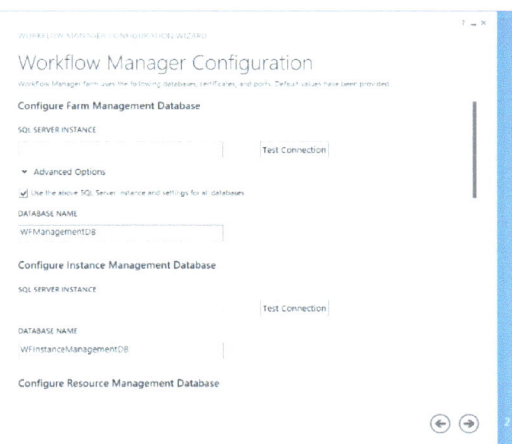

On this section of page 1 enter the SQL Server name and instance and database names for the Workflow Manager databases

You can click on the **Test Connection** button to ensure you can connect to the SQL Server

Service Account

Scrolling down page 1 will reveal more settings

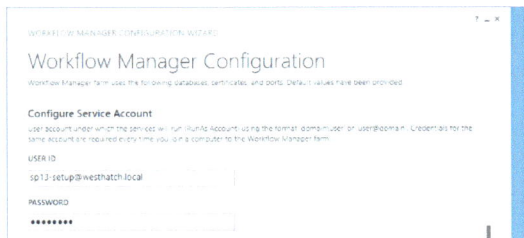

When setting the service account ensure you enter the **USER ID** as a full qualified domain user as shown in the screenshot

Certificates

The next section of page 1 covers the setup of certificates and ports that will be used by Workflow Manager, when you initially scroll down to this section the tick box for auto generation of certificates will be ticked, this will create self-signed certificates, which in a production environment is not a good idea. Remove the tick from the box to assign certificates manually

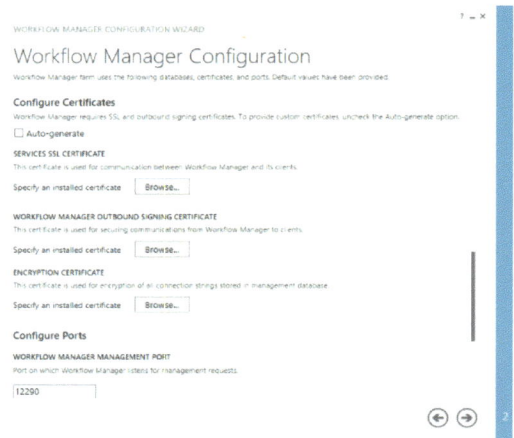

At this point you will need to have a certificate installed onto the Workflow Manager server which you can then use the browse button to find and assign to each service in turn

Port Numbers

The final section of page 1 covers the ports used by Workflow Manager, by default the port numbers are 12290 for HTTPS and 12291 for HTTP

However http is not enabled by default and if you are deploying Workflow Manager in a production environment it is not recommended to enable it

If you require http then simply click the box next to **Allow Workflow management over HTTP on this computer**

Service Bus

Page 2 of the configuration wizard configures the settings for the Service Bus

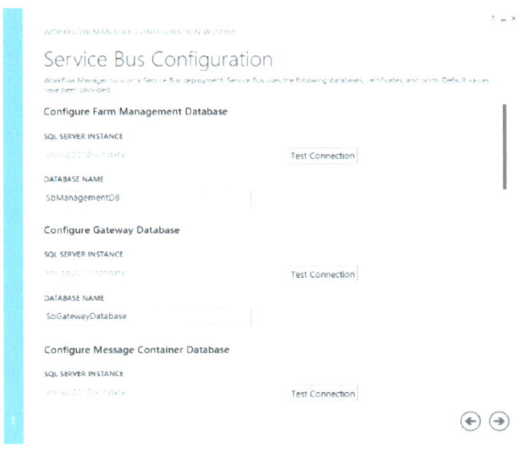

On page 2 you can set the SQL database names (the SQL server instance name is pre-populated from the settings on page 1) and also the settings for service accounts and certificates

You can tick boxes that set the service bus to use the same settings as page 1 for the service account ID

You will also need to choose the certificates in the same way as you chose them on page 1

The final summary screen shows you the settings that will be used to configure Workflow Manager

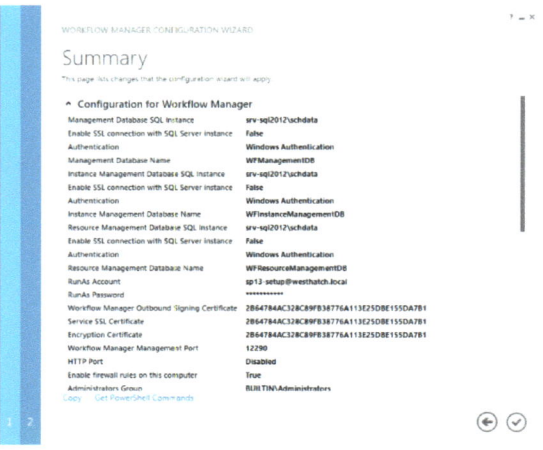

Click the tick to complete configuration, a pop up window will show you the progress of the configuration

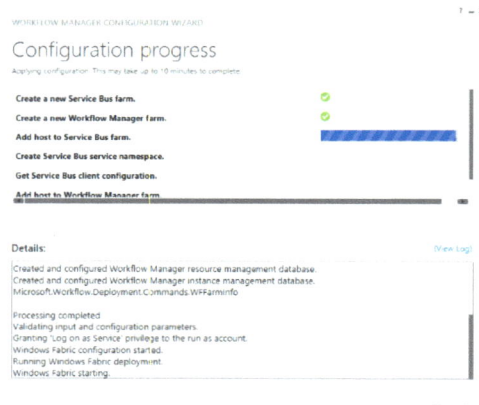

Connecting Workflow Manager To SharePoint 2013

There are 4 options open to you once you have installed the Workflow Manager and they depend on two things:

- Did you install as a standalone Server
- Are you using HTTP or HTTPS for communication

So there is scope for 4 different scenarios

Standalone Workflow Manager Server communicating with HTTP

Standalone Workflow Manager Server communicating with HTTPS

Collocated Workflow Manager Server communicating with HTTP

Collocated Workflow Manager Server communicating with HTTPS

In this example the Workflow Manager is a standalone server and going to use HTTPS to communicate to the SharePoint farm so here are the steps to configure that scenario

Install Workflow Manager Client

The Workflow Manager client is required on all SharePoint 2013 web front ends when the Workflow Manager Server is a standalone server

Logon to each SharePoint 2013 web front end and install the Workflow Manager Client, this can be downloaded from http://go.microsoft.com/fwlink/p/?LinkID=268376

Use the downloaded file to install the client, it will load up the Web Platform Installer, simply follow the steps to install the client

Connection Using PowerShell

Run the SharePoint Management Shell as an administrator

You now need to register the Workflow Service to a site collection using the following command

Register-SPWorkflowService –SPSite
"http://myserver/mysitecollection" –
WorkflowHostUri
"http://workflow.example.com:12290"

Registration of the Workflow service uses the same PowerShell command whether you have a standalone or collocated Workflow Manager Server, the only difference is that the client is installed by default when you install the Workflow Manager Server and so does not need to be installed separately when running in collocated scenario

Testing The Configuration

Your Workflow Manager server should now be associated with your SharePoint 2013 site

To test that we can now access SharePoint 2013 workflows we need to download SharePoint Designer 2013 from the Microsoft website, this link will access the download site
http://www.microsoft.com/en-us/download/details.aspx?id=35491

After installing SharePoint Designer 2013, load it up and click on Open Site

Navigate to the site collection you connected to the Workflow Manager Server and click **Open**

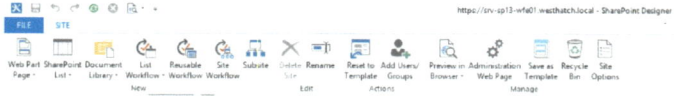

From the ribbon select **List Workflow** and select **Documents**

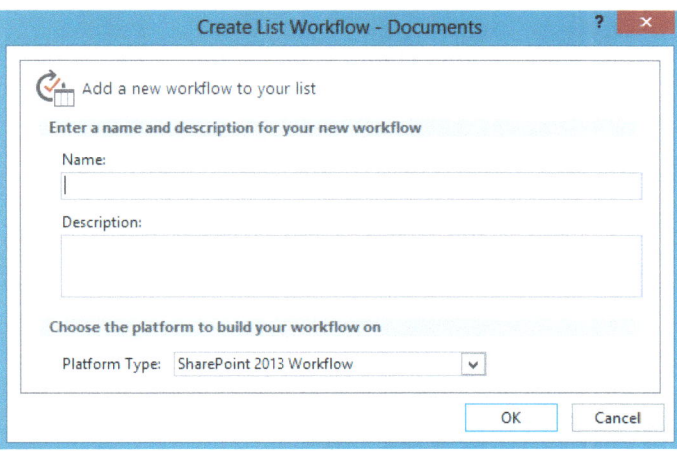

If SharePoint 2013 Workflow is selectable in the drop down then the connection was successful and you can continue to create new expressive workflows

ABOUT THE AUTHOR

Alan Richards has been working in the IT industry for over 20 years and during that time has been at the forefront of using IT. He has led teams that have been among the first to roll out Windows, Exchange and SharePoint, many of these successes have been showcased in Microsoft case studies. Recently Alan's work has concentrated on SharePoint & Office 365 and the implementation of these technologies in organisations to enhance business processes and efficiency. Alan is also a regular speaker & blogger and has been a SharePoint MVP since July 2011.